I0469348

I-75 Multiple Vehicle Collision/
Mass Casualty Incident
Collier County, Florida

Investigated by: John Lee Cook, Jr.

This is Report 155 of Major Fires Investigation Project conducted by Varley-Campbell and Associates, Inc./TriData Corporation under contract EME-97-CO-0506 to the United States Fire Administration, Federal Emergency Management Agency.

Homeland Security

Department of Homeland Security
United States Fire Administration
National Fire Data Center

U.S. Fire Administration Fire Investigations Program

The U.S. Fire Administration develops reports on selected major fires throughout the country. The fires usually involve multiple deaths or a large loss of property. But the primary criterion for deciding to do a report is whether it will result in significant "lessons learned." In some cases these lessons bring to light new knowledge about fire--the effect of building construction or contents, human behavior in fire, etc. In other cases, the lessons are not new but are serious enough to highlight once again, with yet another fire tragedy report. In some cases, special reports are developed to discuss events, drills, or new technologies which are of interest to the fire service.

The reports are sent to fire magazines and are distributed at National and Regional fire meetings. The International Association of Fire Chiefs assists the USFA in disseminating the findings throughout the fire service. On a continuing basis the reports are available on request from the USFA; announcements of their availability are published widely in fire journals and newsletters.

This body of work provides detailed information on the nature of the fire problem for policymakers who must decide on allocations of resources between fire and other pressing problems, and within the fire service to improve codes and code enforcement, training, public fire education, building technology, and other related areas.

The Fire Administration, which has no regulatory authority, sends an experienced fire investigator into a community after a major incident only after having conferred with the local fire authorities to insure that the assistance and presence of the USFA would be supportive and would in no way interfere with any review of the incident they are themselves conducting. The intent is not to arrive during the event or even immediately after, but rather after the dust settles, so that a complete and objective review of all the important aspects of the incident can be made. Local authorities review the USFA's report while it is in draft. The USFA investigator or team is available to local authorities should they wish to request technical assistance for their own investigation.

This report and its recommendations were developed by USFA staff and by TriData Corporation, Arlington, Virginia, its staff and consultants, who are under contract to assist the Fire Administration in carrying out the Fire Reports Program.

The U.S. Fire Administration greatly appreciates the cooperation received from the Florida Highway Patrol; Collier County offices of the Sheriff, Emergency Medical Services, and Emergency Management; and the Golden Gate Fire Control and Rescue District.

For additional copies of this report write to the U.S. Fire Administration, 16825 South Seton Avenue, Emmitsburg, Maryland 21727. The report is available on the Administration's Web site at http://www.usfa.dhs.gov/

U.S. Fire Administration

Mission Statement

As an entity of the Department of Homeland Security, the mission of the USFA is to reduce life and economic losses due to fire and related emergencies, through leadership, advocacy, coordination, and support.

We serve the Nation independently, in coordination with other Federal agencies, and in partnership with fire protection and emergency service communities. With a commitment to excellence, we provide public education, training, technology, and data initiatives.

ACKNOWLEDGMENTS

The Federal Emergency Management Agency, U.S. Fire Administration gratefully acknowledges the cooperation of the members of the Florida Highway Patrol; Collier County Emergency Management Agency, Sheriff's Department, and Emergency Medical Services; Golden Gate Fire Control and Rescue District; the Collier County Medical Examiner. Every one who assisted in the preparation of this report was generous with his or her time, expertise, and counsel.

TABLE OF CONTENTS

OVERVIEW 1

KEY ISSUES 2

THE COMMUNITY 3

THE HIGHWAY 3

THE INCIDENT 3

INVESTIGATION 6

LESSONS LEARNED 7

APPENDIX A – PHOTOGRAPHS 9

APPENDIX B – MAPS AND DIAGRAMS 23

I-75 Multiple Vehicle Collision/Mass Casualty Incident
Collier County, Florida
January 25, 2002

Investigated By: John Lee Cook, Jr.

Local Contacts: Paul D. France, Lieutenant
Naples Sub-District Commander
Florida Highway Patrol
3205 Beck Boulevard
Naples, Florida 34114
(941) 354-2377

Marta U. Coburn, M.D.
District Medical Examiner
3838 Domestic Avenue
Naples, Florida 34104
(239) 434-5020

Kenneth F. Pineau
Emergency Management Director
Collier County Government
3301 East Tamiami Trail
Naples, Florida 32112
(941) 774-8444

Pat Mullen, Captain
Collier County Sheriff's Office
(941) 793-9204

Dan Bowman, Captain
Emergency Medical Services
   Department
(941) 774-8459

Donald R. Peterson, Fire Chief
Golden Gate Fire Control and
   Rescue District
4741 Golden Gate Parkway
Golden Gate, Florida 34116
(941) 455-2121

OVERVIEW

Dense fog on the morning of Friday, January 25, 2002 contributed to a massive motor vehicle collision along I-75 in a rural area of Collier County, Florida. Three adult males, ranging in age from 24 to 49, were killed and 13 people were injured. All three men died immediately from blunt force trauma, according to the Medical Examiner, and had to be extricated from their respective vehicles.

Access to the site was limited by both its remote location and the massive traffic backups, which ensued following the multiple collisions along the Interstate. Weather conditions precluded the use of helicopters. Therefore, five ambulances from the Collier County EMS were used to transport all of the victims to Naples for treatment. None of the injuries proved to be life threatening and the patients were released from the hospital. Six victims were taken to the Community Hospital and another seven patients were transported to the Cleveland Clinic.

1

The incident occurred between mile markers #85 and #86 on I-75 in an area known as Alligator Alley, which is the major east/west thoroughfare in southern Florida. A total of 27 vehicles, including 17 tractor-trailer trucks, were involved in 8 separate collisions, which were spread out over a distance of 1/2 mile. The incident, which occurred just before 05:00 hours, shut down I-75 in both directions for approximately 8-1/2 hours literally backing up traffic from the Gulf of Mexico to the Atlantic Ocean. Eastbound lanes were not opened until 14:30 hours at 18:15 hours both westbound lanes were finally opened for traffic.

It took 58 emergency responders to manage the incident, including units from the Florida Highway Patrol, Collier County Sheriff's Department, Collier County EMS, and the Golden Gate Fire Control and Rescue District. Rescuers were assisted at the scene by personnel from the County Medical Examiner's Office, the Florida Department of Transportation, and the Red Cross.

Following the removal of the dead and injured, a massive cleanup effort was required to restore traffic in both directions. Inmates from the Hendry Correctional Institution were bused in to help in the effort, which required the services of a number of heavy-duty wreckers, front-end loaders, and dumpsters. Due to the damage sustained by a number of the trucks, cargos had to be transferred to other vehicles in order to be removed from the scene.

KEY ISSUES

Issues	Comments
Access & Travel Distance	The incident occurred in a remote area, blocking the main traffic artery in both directions. Heavy traffic congestion hampered emergency responders and there were very few alternative routes to divert traffic away from the area. Fog during the early phases of the incident also precluded the use of aircraft. The closest emergency responders were 40 minutes away from the site and due to the terrain; backup units were farther away than may normally be encountered in a suburban or urban environment, which delayed the efforts to resolve the incident.
Communications	Communications at the incident site were a major issue due to the non-operability of the radio systems of the several agencies involved in the incident. System interoperability is a common deficiency at almost all major events.
Logistics	The need to remove a large number of vehicles from the highway, clear debris, and salvage and transfer cargos poses significant logistical issues that may challenge most jurisdictions. Such issues should be considered during the emergency planning process.
Resources	The scope and complexity of the incident required considerable commitment of human and material resources. Multiple fatalities and injuries also place a burden on ancillary resources such as hospitals and the morgue. In some instances, it may be necessary to transport those persons stranded at such an incident away from the scene and to temporarily provide them with shelter. Planning is the key to successfully managing this type of event.
Time of Day	The incident occurred at approximately 05:00 hours. Had the incident occurred later during the height of the rush hour the number of potential victims and vehicle involved could well have been significantly higher.
Weather	Temperatures ranged from the low 60's to a high of 81 degrees Fahrenheit on the 25th and, apart from the fog, the skies were generally clear with no precipitation. Winds were calm. Florida rarely suffers from extreme cold, but the summer can produce high temperatures and humidity, which includes significant thunderstorm activity. Had the incident occurred during the summer, the weather would potentially have had an impact on both victims and rescuers. Provisions would also have also been necessary to shelter the large number of motorists stranded in their vehicles on the highway as well, particularly the very old and the very young. Ample hydration would have been paramount.

THE COMMUNITY

Collier County was created by an act of the State Legislature on May 8, 1923 and is located in southwest Florida on the Gulf Coast. Directly west of Miami and 35 miles south of Fort Myers, Collier County is the largest of Florida's 67 counties and has a population of 265,769. At 2,305 square miles, the County is larger than the state of Delaware (See Map #1). Naples is the county seat.

Collier County provides emergency medical services to the entire county, which includes air transportation capabilities. Fire protection is provided by a number of municipalities and fire protection districts. The Collier County Sheriff's Department provides law enforcement to the unincorporated portions of the County as well as managing the 9-1-1 emergency communications center.

THE HIGHWAY

Alligator Alley, which is also known as State Road 93 and/or Interstate 75, is a four-lane divided highway that is oriented in an east west direction Constructed of smooth, worn asphalt, the roadway is straight and level with a zero percent grade. Each of the two eastbound lanes is 12 feet in width. The lanes are divided by a broken white painted line with reflective delineator buttons and the outside lane is bordered by 11-foot wide emergency lane or apron that leads to a 30-foot wide grass shoulder that slopes away from the highway for drainage purposes. The inside lane is bordered by a 5-foot wide paved apron, which leads to a grass covered median. Approximately 80 feet in width, the median slopes from both the east and westbound lanes to the center for drainage. Both the inside and outside lanes are separated from their respective aprons by a solid white painted line.

The two westbound lanes are also constructed of asphalt with the inside lane measuring 11 feet, 6 inches in width and the outside lane measuring 12 feet in width. Travel lanes are separated by a broken, black and white painted centerline and there is a paved apron measuring 4 feet, 6 inches in width adjacent to the inside lane. The apron and inside lane are separated by a solid, yellow painted line. Likewise, a solid, painted white line separates the outside lane from a paved emergency lane, which is 10 feet, 6 inches in width. A 42-foot, 6-inch wide shoulder borders the outside lane and slopes away from the roadway for drainage purposes.

The maximum posted speed limit in either direction is 70 mph. There is a speed limit signpost on the south shoulder of the eastbound lane 0.1 miles west of the crash site and another sign is posted 2.3 miles east of the crash site for westbound traffic. There is no overhead lighting, guardrails, or obstructions in the area.

Alligator Alley is very isolated, with limited access and is the major east-west route in southern Florida. The only other east-west route in this portion of the state is US 41. The roadway is bordered on the north by the Florida Panther National Wildlife Refuge and by the Fakahatchee Strand State Preserve on the south.

THE INCIDENT

During the early morning hours of Friday, January 25, 2002, temperatures in the Naples area dipped into the low 60's while the relative humidity rose to approximately 94 percent. Around 02:00 hours, fog began to form. The National Weather Service, stationed at the Naples Municipal Airport, issued a hazardous weather outlook at 04:30 hours for Alligator Alley and Tamiami Trail. The advisory indicated that the lower elevations in the region might experience widespread, dense, patchy fog.

As foggy conditions developed along I-75, the driver of a pickup truck traveling westbound to Naples stopped his truck due to poor visibility. He believed that he had parked on the shoulder of the outside lane. His pickup, however, protruded into the traffic lane and was rear-ended by a tractor-trailer truck transporting automobiles. The truck driver stated that he came upon the stopped vehicle, which had been obscured by the dense fog, and had swerved to miss the pickup, but collided with the right rear corner of the pickup and also hit a car. The car was forced into the eastbound lanes of traffic, which resulted in another collision.

Before the incident would be concluded; 27 vehicles, including 17 tractor-trailer rigs, would become tangled in 8 separate pileups (6 westbound and 2 eastbound) along a 1/2-mile stretch of I-75, blocking traffic in both directions. Thirteen people would be injured and three lives would be lost. It would take more than 13 hours to restore conditions to normal.

All of this occurred between mile markers 85 and 86, approximately 9 miles west of State Road 29 and 15 miles east of Collier Boulevard (951). The highway is both level and straight at that point and is four-lane divided expressway without service roads.

At 04:59 hours; after receiving numerous reports via cellular telephones, the Collier County 9-1-1 center dispatched units from the Sheriff's Office, the Collier County EMS, and the Golden Gate Fire Control and Rescue District to the collision. The communications center serves as the 9-1-1 answering point for the entire County and dispatches (800 MHz) all of the law enforcement and fire agencies in the County. Fire and EMS agencies are typically dispatched on separate tactical channels, but move to a common channel for joint operations. The Communications Center also notified the Florida Highway Patrol, which is dispatched (VHF) out of Fort Myers. The Highway Patrol is in command on Interstate Highways.

Emergency responders did not have a clear understanding of the incident when they initially responded because of the many conflicting reports still being received by the 9-1-1 Center. At least one caller advised that placarded vehicles, which would indicate the presence of hazardous materials, were involved. That is principle reason that the fire department was included on the initial dispatch. It was later determined that the placards had moved as a result of the impact of collision, which resulted in the display of a radioactive placard on one truck and an explosives placard on another. The only hazardous materials involved in the incident were approximately 35 gallons of diesel that spilled in the median when a saddle tank on an 18-wheeler ruptured.

Responders thought they were all going to a single, but common incident. Visibility was so poor that emergency responders had no warning that they were near the incident until they were right on top or it. The high volume of 9-1-1 calls reporting the incident also resulted in conflicting mile markers being given to responders. An unconfirmed report of a controlled burn was also made. A number of responders indicated that they smelled smoke in the vicinity of the incident. The addition of smoke to the fog may have further limited visibility at the scene.

Units from the Sheriff's Department were the first to arrive on the scene. They went on location at the eastbound incident; but could not see very well due to fog. Visibility at times was limited to approximately ten feet. Deputies soon realized that there were collisions in both directions, but were not sure at the time of the magnitude of the incident. They realized that a lot of activity was going on around them that they could not see because they were hearing sounds of additional crashes as they occurred as well as the calls from the injured. It was not until daylight (07:15 hours) and the fog began to clear up that they began to realize the full extent of the incident.

Upon their arrival, Fire and EMS personnel established a command post on the west end of the incident. A staging area was also established and the EMS established a casualty collection point near the command post. Fire/EMS personnel triaged the area and searched for victims. This proved to be difficult, however, because most of the occupants who were able to do so had abandoned their vehicles in order to get off highway. They gathered along the fences, which ran along both sides of the highway. There was also a concern that there might be victims in the nearby canal, which is a common occurrence, but this was not the case.

Chaos and confusion was ever present, compounded by both the darkness and the fog. People were walking about all over and some vehicles were turning around on the highway in an attempt to flee the congestion that was constantly increasing. Additionally, law enforcement agencies do not participate in the local Fire and EMS Incident Management System. Therefore, no one from a law enforcement agency was in the command post, which made a variety of efforts more difficult.

Rescuers determined that 13 individuals required transportation to a medical facility, primarily due to cuts and back or neck pain. Seven patients were taken to the Cleveland Clinic in Naples and another six patients to Naples Community Hospital. None of the injured proved to be critical and all of them were treated and released. Patients were transported by ground by one of five ambulances used in this effort because air operations had been suspended due to the fog.

Most of the injured thought that they were okay when they were initially triaged by EMS personnel, but as their excitement subsided many realized that they were injured. EMS personnel also speculated that at least some of the patients realized that transportation by an ambulance was the only way to leave scene so they opted to be transported as a means of escaping the scene.

The Lee County trauma center was initially placed on standby until the extent and nature of the injuries could be determined. Since there were only minor injuries, the number of patients did not create any significant problems at the local emergency rooms. If there had been many more injuries or had a number of the injuries been more severe, the patient load would have taxed the ability of the hospital facilities in Naples to treat all of the patients.

Three people died as a result of injuries sustained in two of the collisions. A Hispanic male, 49, was killed in the westbound lane when the 1993 Mack concrete powder truck he was driving hit the rear of another truck that was hauling produce. Two brothers, also Hispanic males aged 24 and 30, were killed in a five-vehicle pileup in the eastbound lane when their 1998 Dodge van struck the rear end of a truck hauling long wooden poles. The truck had stopped due to a previous collision, which was blocking the road. All three victims had to be extricated. It took firefighter 15 minutes to extricate the two victims in the van and 50 minutes to extricate the truck driver.

The medical examiner was requested to respond at 07:56 hours to remove the three victims' bodies. A private contractor transported the bodies to the morgue for autopsy. The medical examiner ruled that all of the victims were killed instantly and died as a result of multiple blunt force trauma injuries.

At least 58 emergency responders were involved in the resolution of the incident. There were 9 officers from the Florida Highway Patrol, 20 deputies from the Sheriff's Department, 16 firefighters, 10 EMS personnel, and three personnel from the Florida Department of Transportation.

Personnel from the Florida Department of Environmental Protection also responded and managed the 35 gallons of spilled diesel fuel, which was confined to the median. The contaminated soil was removed.

The County Emergency Management Agency was notified of the incident, but the isolated nature of the incident did not warrant activation of the County's Emergency Operations Center. Local officials notified the Red Cross and the "warning point" at the Florida Department of Emergency Management.

Law enforcement personnel were not only used at the scene, but also to control traffic distant from the incident site. They stopped all eastbound traffic at the toll plaza in Fort Lauderdale as well at the intersections of Highways 951 and 29 with I-75.

The Red Cross established a rehab sector and brought portable toilets, water, and food to scene for the emergency responders and for the many travelers stranded at the scene.

Three members of the Sheriff Department's critical incident stress management team responded to the scene to assist emergency responders due to severity of the trauma to the deceased. The team also helped with the families of the deceased and injured as well as many others that had been stranded by accident.

The Sheriff's Department also sent their victim's advocate team to the scene. They helped to de-escalate the situation, which had become tense and hostile. A number of truckers and other drivers were very angry, particularly at one trucker that was estimated to have been traveling at 70 mph when he slammed his rig into four other trucks, almost killing and injuring a number of people.

As news of the incident began to spread, a number of insurance adjusters congregated at the scene to begin to process insurance claims. The media also responded from both sides of the incident. Once the fog cleared, they used their helicopters to take photos of the site.

Given the importance of I-75 as a major east-west transportation route, the objective was to reopen the roadway as quickly as possible. While it is the carrier's responsibility to secure and cleanup their cargo, the magnitude of the incident proved to be problematic because 17 18-wheelers were involved. Approximately 10 to 12 of the vehicles had to be towed from the site, which meant that their loads had to be off loaded or if spilled, salvaged from the surrounding debris.

Removing all of the trucks that were blocking the roadway present a problem as well, due to the availability of a sufficient number of large wreckers needed to move the trucks. Wreckers from nearby counties were called to the scene and private contractors responded with two Bobcats and dumpsters to assist in the recovery efforts. A sergeant from the Florida Highway Patrol was assigned to stage and utilize the responding wreckers.

Inmates from the nearby Hendry Correctional Institution were bused to the scene and were pressed into service to help unload the damaged trucks and load the cargo onto trucks brought to the scene to recover the damaged property. Boxes of produce, Sunkist Orange Soda, and A&W Diet Root Beer lined the median of the highway and a Canada Dry Truck, among others, had to be off loaded.

I-75 was reopened to eastbound traffic at 14:30 hours. At 16:00 hours one of the westbound lanes was opened and the entire roadway was fully opened at 18:00 hours, approximately 13 hours after the incident occurred.

INVESTIGATION

Jurisdiction at the crash site belonged to the Florida Highway Patrol. They were assisted in their investigation by the Motor Compliance Division of the Florida Department of Transportation, which is mandated to investigate incidents that involve commercial vehicles. The agency's charge is to determine if there are violations that contribute to the cause of such incidents. Officers from the Collier county Sheriff's Department also assisted with the investigation.

The investigation determined that the traffic control devices on the highway were adequate and did not contribute to the collision, nor were there any visual obstructions or road defects in the area of the collision. Ambient conditions were, however, a factor. The heavy fog made driving conditions extremely difficult and severely limited visibility in both directions.

Local authorities stated that they felt like the number of fatalities should have been higher. For example the driver of the Canada Dry truck was in front of his truck on a cellular telephone when the truck was struck by another vehicle and he barely escaped injury.

The driver of the second 18-wheeler in the westbound lane (See Diagram #4) was ticketed with a uniform traffic citation because of portion of his vehicle was in the roadway when he pulled off the road, which resulted in the first of the westbound collisions. The driver was tested for alcohol and controlled substances, but none were detected.

Two occupants of the van in the eastbound lane sustained fatal injuries and were found to have not been wearing the occupant restraint system at the time of the collision, which is a violation of Florida Statutes. Investigators concluded that the driver of the van was the sole cause of the property damage and deaths in the collision and the Medical Examiner's Officer determined that no drugs or alcohol were present in the deceased driver of the van.

A Lake Park, Florida man was also cited for careless driving and driving with a suspended license.

LESSONS LEARNED

1. **Pre-Incident Planning is Crucial:**

   Collier County maintains an up to date Emergency Operations Plan, which is regularly exercised. Although, the size of the incident did not necessitate the activation of the plan or the Emergency Operation Center, the applicable portions of the plan were successfully implemented by the responding agencies. Most of the emergency responders knew each other as a result of the planning and exercising activities and this allowed the incident to be resolved with a minimum of difficulty.

2. **Unity of Command is Essential:**

   Fire and emergency medical agencies are familiar with and routinely use an incident management system. Law enforcement agencies, however, do not routinely use a similar incident management system and as a result, key law enforcement officials were not present in the fire department command post at this incident. The incident, however, reinforced the need for all emergency responders to use a unified system of command, particularly during the early phases of an incident when conditions tend to be chaotic at best. Everyone has a responsibility at an incident, which can be more effectively coordinated by a centralized, unified command.

   Unified command and effective incident management also reinforces the need for strict accountability of personnel operating at the scene. When there are multiple agencies (in this case the Sheriff's Office and Highway Patrol) operating at an incident that do not participate in command and accountability activities, it is impossible to know how many emergency responders are on the scene and where they all are deployed.

3. Incident Debriefing:

Participants expressed the need to conduct a debriefing following an incident of this magnitude in order to build upon the successes of the incident and improve upon the deficiencies, which might have been identified through a formal debriefing process. Such sessions reinforce the value of planning and the utility of an incident management system. Equally important is the need to provide critical incident stress debriefing. Responders to this incident treated multiple injuries, extricated three fatalities with severe trauma, and experienced the fear of hearing collisions occur all around them without being able to adequately see if they needed to move to a place of safety.

4. Do Not Demobilize Too Quickly:

Once the patients were evacuated to medical facilities, EMS was released from the scene. The cleanup, however, took many more hours to complete. Due to the amount of the wreckage and the nature of the debris, it would have been prudent to maintain an EMS crew on standby to treat anyone that might potentially have been injured or have been overcome by climatic conditions. Additionally, a safety officer should have been in place for the duration of the incident.

5. Resource Management is Critical:

The incident was at least 40 minutes from the nearest emergency service agency. Nevertheless, the scope and magnitude of the incident required a significant commitment of personnel and equipment from law enforcement, fire, and EMS agencies. The cleanup was significant and very few jurisdictions have access to the large number of heavy duty wreckers required to remove all of the large trucks involved in the incident. Prior planning for such an eventuality is essential to ensure that adequate resources will be available, when and if needed.

Cargo and other debris was scattered throughout the site. Fortunately, none of it involved hazardous or dangerous materials. Even though the logistics of handling the loads remain the responsibility of the individual carriers, thought should be given to identifying a potential labor pool to assist in an incident of this type. Minimally, it will be necessary to clear the roadway of debris, and there is an issue of security if cargo and other valuables are going to remain on the side of the road for an extended period of time. Fortunately, a nearby correctional institution was able to assist in this incident, but such a resource may not always be available in other jurisdictions.

6. Traffic Management:

Alligator Alley is the major east-west highway in southern Florida and the incident caused traffic to be stopped for hours. Vehicles were literally backed up from the Gulf of Mexico to the Atlantic Ocean. Very few alternative routes were available for the diversion of the traffic, which further complicated the situation. The local media assisted throughout the incident by providing updated traffic reports on both radio and television.

Many of the motorists were from out of state and were not familiar with the area. Some of them became lost and disoriented, adding to the congestion. The Sheriff's Office responded quickly by producing a large number of copies of maps of alternate routes that were handed out at the control and diversion points. The maps helped manage the flow of traffic and eased congestion somewhat. During the planning process, emergency managers might give some thought to following their example should a similar incident occur in their jurisdiction.

APPENDIX A

Photographs

Photo	Description	Source
1	Florida Highway patrolman investigates van in which two men were killed when their van rear-ended a tractor trailer truck hauling wood in an eastbound lane of Alligator Alley	Naples Daily News: Gary Coronado photographer
2	Aerial view of westbound lanes of Alligator Alley	Naples Daily News: Lisa Krantz photographer
3	Firefighters extricate the body of the driver of a tractor-trailer driver killed in the westbound lane of Alligator Alley	Naples Daily News: Lisa Krantz photographer
4	Close-up of the cab of the concrete truck in which the driver was killed	Collier County SO
5	Debris spilled during collision	Collier County SO
6	Numerous tractor trailer trucks involved in the incident	Collier County SO
7	Ruptured diesel fuel tank	Collier County SO
8	Additional view of tractor trailer rigs	Collier County SO
9	Pickup that caused the original collision to occur	Collier County SO
10	Another view of the concrete truck in which the driver was killed	Collier County SO
11	Congestion at the scene looking from the road side toward the median	Collier County SO
12	Example of the degree of damage	Collier County SO
13	Extrication efforts by firefighters to remove the body of the victim from the concrete truck	Collier County SO

1. Florida Highway patrolman investigates van in which two men were killed
when their van rear-ended a tractor trailer truck hauling wood
in an eastbound lane of Alligator Alley

2. Aerial view of westbound lanes of Alligator Alley

3. Firefighters extricate the body of the driver of a tractor-trailer driver killed in the westbound lane of Alligator Alley

4. Close-up of the cab of the concrete truck in which the driver was killed

5. Debris spilled during collision

6. Numerous tractor trailer trucks involved in the incident

7. Ruptured diesel fuel tank

8. Additional view of tractor trailer rigs

9. Pickup that caused the original collision to occur

10. Another view of the concrete truck in which the driver was killed

11. Congestion at the scene looking from the road side toward the median

12. Example of the degree of damage

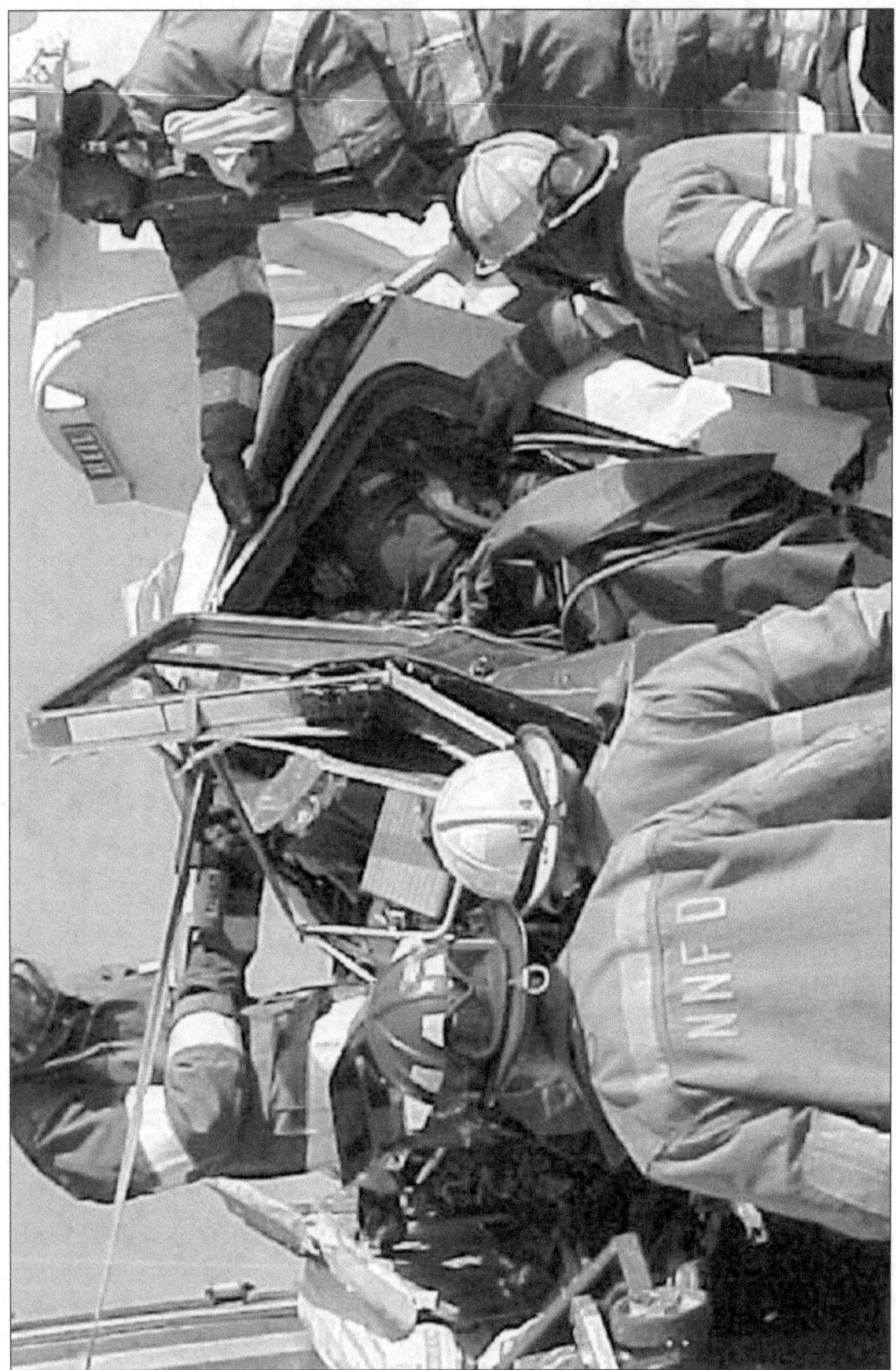

13. Extrication efforts by firefighters to remove the body of the victim from the concrete truck

APPENDIX B

Maps and Diagrams

Map #1:   Map of Collier County
          Source: County Web site

Map #2:   Map of the Accident Site
          Source: Naples Daily News

23

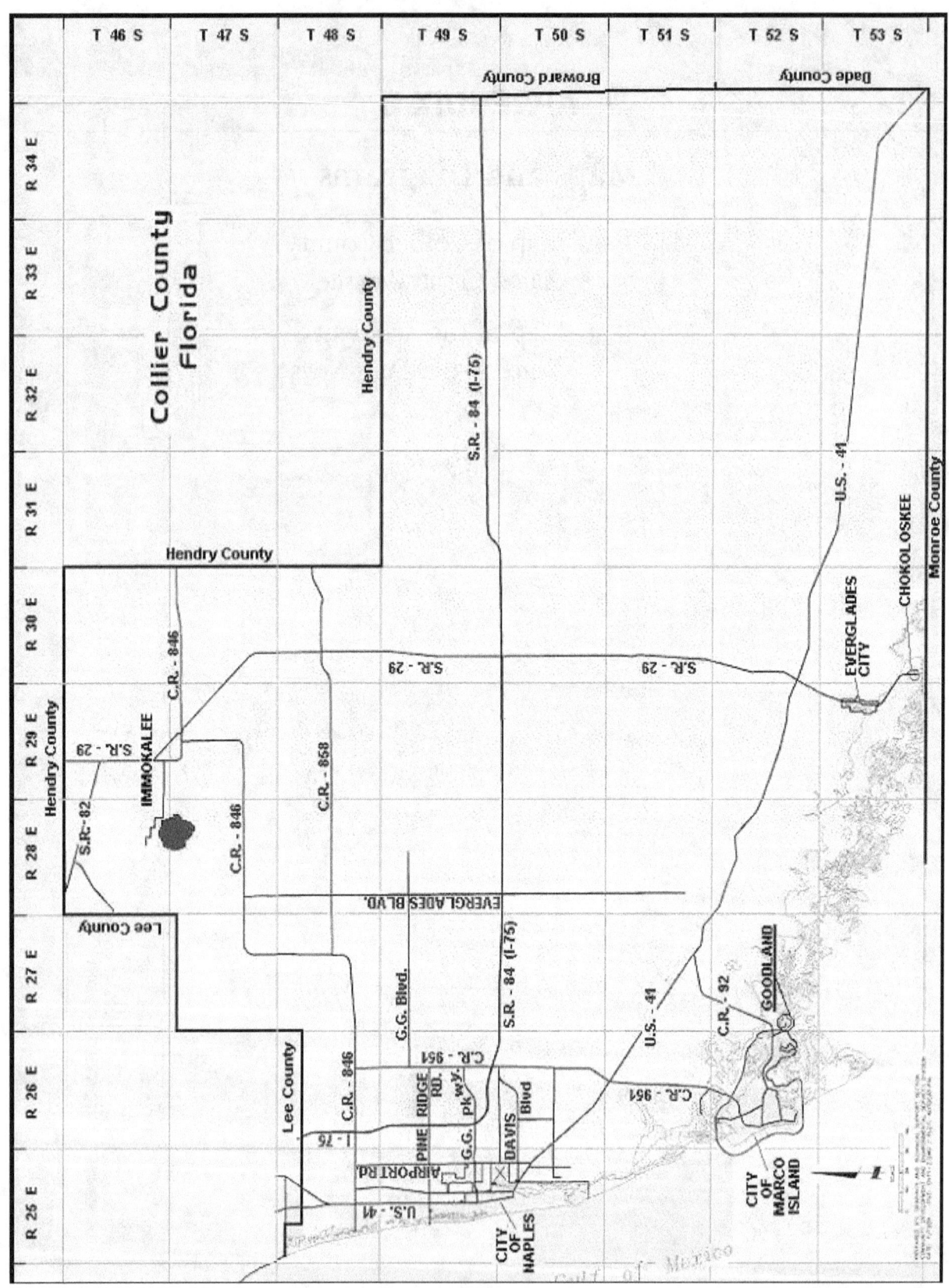

Map #1: Map of Collier County

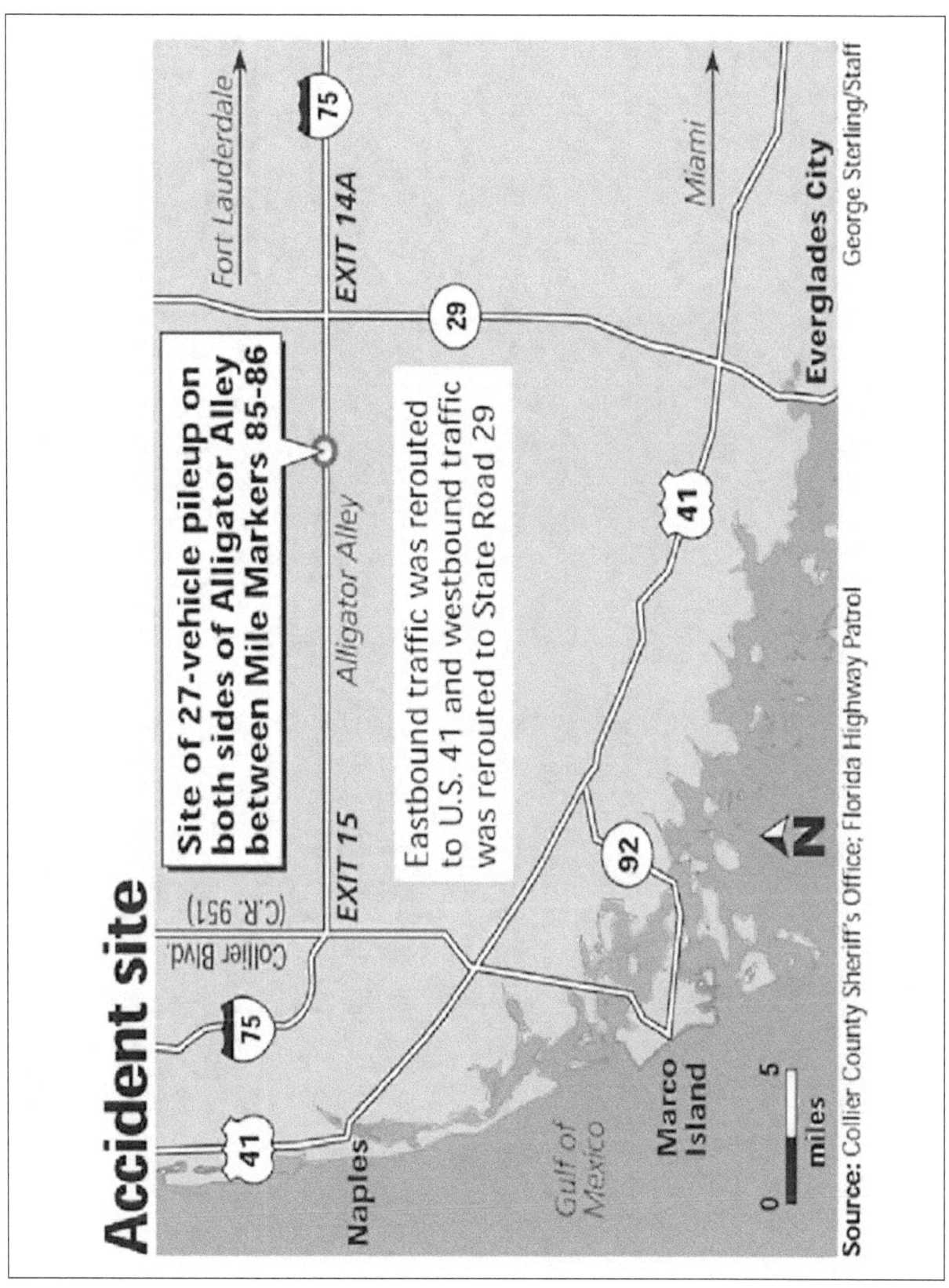

Map #2: Map of the Accident Site

www.ingramcontent.com/pod-product-compliance
Lightning Source LLC
Chambersburg PA
CBHW081244170526
45165CB00009B/3194